بسم الله الرحمن الرحيم

ENGINEERING MATHEMATICS

Mathematical Modeling, Differential Equation Principles, and Applications (2024)

DAVOOD SHADMANI

Copyright © 2024 by Davood Shadmani.

All rights reserved. No part of this book may be used or reproduced in any form whatsoever without written permission except in the case of brief quotations in critical articles or reviews.

Printed in the United States of America.

For more information, or to book an event, contact:

davoodshamani1978@gmail.com

Cover design by Davood Shadmani

Second Edition: October 2024

CONTENTS

- INTRODUCTION 9
  - UNDERSTANDING EQUATIONS AND THE BASIC STEPS FOR BUILDING EQUATIONS 9
  - STEP 1: DEFINING THE PROBLEM 11
  - STEP 2: IDENTIFYING THE MEASURABLE PROPERTIES: A CONTROL VOLUME PERSPECTIVE 12
    - CONTROL VOLUME (INTENSIVE AND EXTENSIVE PROPERTIES) 12
    - SURROUNDING SPACE 13
    - SURROUNDING TIME 14
    - BEHAVIOR 15
    - CONDITION 16
  - NOTES ON CLASSIFICATION 17
  - STEP 3: SPECIFYING THE REFERENCE IDENTITY 18
  - STEP 4: ANALYZING ENERGY INTERACTIONS 20
  - STEP 5: FINALIZING THE OUTCOME EQUATION 21
  - STEP 6: MODIFYING THE EQUATION 21
  - APPLICATIONS 22
- CHAPTER 1 23
- CATEGORY A 23
  - ENERGY, THE REFERENCE QUANTITY 23
  - STEP 1: DEFINING THE PROBLEM 24
  - STEP 2: IDENTIFYING MEASURABLE PROPERTIES 26
  - STEP 3: IDENTIFYING THE REFERENCE IDENTITY 29
  - STEP 4: ANALYZING ENERGY INTERACTIONS 29
  - STEP 5: FINALIZING THE OUTCOME EQUATION 32
  - STEP 6: MODIFYING THE EQUATION 33
- CHAPTER 2 36
- CATEGORY B 36

- STEP 1: DEFINING THE PROBLEM 36
- STEP 2: IDENTIFYING MEASURABLE PROPERTIES 37
- STEP 3: IDENTIFYING THE REFERENCE IDENTITY 39
- STEP 4: ANALYZING ENERGY INTERACTIONS 40
- STEP 5: FINALIZING THE OUTCOME DIFFERENTIAL EQUATION 42
- STEP 6: DOCUMENT THE OBTAINED DIFFERENTIAL EQUATION 45

CHAPTER 3 47
CATEGORY C 47
- STEP 1: DEFINING THE PROBLEM 47
- STEP 2: IDENTIFYING MEASURABLE PROPERTIES 48
- STEP 3: IDENTIFYING THE REFERENCE IDENTITY 50
- STEP 4: ANALYZING ENERGY INTERACTIONS 51
- STEP 5: FINALIZING THE OUTCOME EQUATION 53
- STEP 6: DOCUMENT EQUATION AND COMMENT 54

CHAPTER 4 55
MATHEMATICAL MODELING SOME PROBLEMS 55
- MATHEMATICAL MODELING OF STEADY-STATE FLOW EQUSTION 56
  - INTRODUCTION TO STEADY-STATE FLOW PROCESS 56
  - PROBLEM 58
  - SOLUTION 59
  - STEP 1: DEFINING THE PROBLEM 59
  - STEP 2: IDENTIFYING MEASURABLE PROPERTIES 61
  - STEP 3: IDENTIFYING THE REFERENCE IDENTITY 64
  - STEP 4: ANALYZING ENERGY INTERACTIONS 66
  - STEP 5: FINALIZING THE OUTCOME EQUATION 68
  - STEP 6: MODIFYING AND VERIFYING THE ACCURACY OF THE EQUATION 70

- MATHEMATICAL MODELING OF HEAT TRANSFER EQUATION 76
  - INTRODUCTION TO THERMAL HEAT TRANSFER 76
  - PROBLEM 78
  - SOLUTION 78
  - STEP 1: DEFINING THE PROBLEM 78
  - STEP 2: IDENTIFYING MEASURABLE PROPERTIES 80
  - STEP 3: IDENTIFYING THE REFERENCE IDENTITY 82
  - STEP 4: ANALYZING ENERGY INTERACTIONS 83
  - STEP 5: FINALIZING THE OUTCOME EQUATION 85
  - STEP 6: DOCUMENT THE OBTAINED EQUATION AND COMMENT 86
- MATHEMATICAL MODELING OF SHAFT WORK EQUATION OF A SYSTEM 87
  - INTRODUCTION TO SHAFT WORK OF A SYSTEM 87
  - PROBLEM 88
  - SOLUTION 89
  - STEP 1: DEFINING THE PROBLEM 89
  - STEP 2: IDENTIFYING MEASURABLE PROPERTIES 90
  - STEP 3: IDENTIFYING THE REFERENCE IDENTITY 93
  - STEP 4: ANALYZING ENERGY INTERACTIONS 95
  - STEP 5: FINALIZING THE OUTCOME EQUATION 98
  - STEP 6: DOCUMENT THE OBTAINED EQUATION AND COMMENT 99

CHAPTER 5 100
A HARD PROBLEM 100
- MATHEMATICAL MODELING OF A PARTIAL DIFFERENTIAL EQUATION FOR THE BEHAVIOR OF VISCOUS INCOMPRESSIBLE FLUIDS ON THREE-DIMENSIONAL SPACE 100
  - STEP 1: DEFINING THE PROBLEM 100
  - STEP 2: INCOMPRESSIBILITY AND THE MEASURABLE PROPERTIES 101

- STEP 3: IDENTIFYING THE REFERENCE IDENTITY 105
- STEP 4: ANALYZING ENERGY INTERACTIONS 106
- DERIVING THE NAVIER-STOKES EQUATION: AN ENERGY-BASED APPROACH 107
- STEP 5: FINALIZING THE OUTCOME PARTIAL DIFFERNTIAL EQUATION 114
- STEP 6: DOCUMENT THE OBTAINED EQUATION AND COMMENT 116

BIBLIOGRAPHY 117

ACKNOWLEDGMENTS 118

INTRODUCTION

Understanding Equations and the Basic Steps for Building Equations

In its simplest form, an equation consists of two sides, the left and right sides, connected by an equal (=) sign. The left and right sides of an equation, respectively known as the "expression" and the "value", contain one or more terms and variables that are symbols to represent unknown or changing values. The value, on the other hand, contains known or fixed quantities.

Although constructing equation(s) to address a problem is challenging, following an exceptional step-by-step guide and principles can make it more manageable, engaging, and even more fun.

Identifying the Problem

Identifying the problem is the foundational step in an effective problem-solving model. This requires formulating well-structured questions that elicit relevant answers, providing a solid foundation for subsequent mathematical modeling.

This process typically includes:

1. **Defining the problem:** Clearly articulating the problem by breaking down the problem and dividing it into smaller and manageable components to understand its root causes.

2. **Specifying the Reference Identity:** Identifying a reference substance, a reference quantity, and a corresponding unit as the reference unit for the problem at hand.

3. **Identifying Measurable Properties:** Characterizing the system's control volume, surrounding environment, temporal context, conditions, and behavior to determine relevant measurable quantities.

4. **Analyzing Energy Interactions:** Assessing energy based on its interactions with the identified measurable system characteristics to quantify energy in relation to system-measurable properties.

5. **Finalizing the Outcome Equation:** Establishing a mathematical formula that accurately represents the desired result consistent with the problem's parameters.

6. **Modifying the Equation (if required):** Confirming the equation's accuracy by comparing its results to real-world data or experimental outcomes.

Identifying the problem sets the foundation for finding a suitable solution.

Step 1: Defining the Problem

Problem decomposition is essential to define a problem effectively. This involves breaking down a complex issue into smaller, more manageable sub-problems. This approach helps to:

- **Identify the core issue:** Pinpoint the exact problem you are trying to solve.

- **Simplify complexity:** Make a complex problem easier to understand and analyze.

- **Identify dependencies:** Understand how distinct parts of the problem relate to each other.

The Process is as follows:

- **Describe the Problem:** Clearly articulate what the problem is. Avoid vague or ambiguous language.
- **Break it Down:** Break down the problem into simpler, manageable components.

- **Analyze Components:** Examine each sub-problem independently. Identify its causes, effects, and potential solutions.
- **Identify Relationships:** Understand how the sub-problems connect and interact. Are there dependencies between them?
- **Prioritize:** Determine the order in which to address the sub-problems. Some problems may be more critical than others.

Step 2: Identifying the Measurable Properties: A Control Volume Perspective

Derived properties are properties calculated or inferred from other data. This framework focuses on properties derived within a defined control volume, considering its interactions with the surrounding space and time.

1. **Control Volume (Intensive and Extensive Properties)**

    Control volume: A control volume represents a defined spatial region used for system analysis. Property classification within this region is in terms of intensive and extensive properties.

    ➤ **Intensive Property:** A substance property that does not depend on the amount of the reference substance (present matter).

- **Extensive property:** A substance property that depends on the amount of the reference substance (present matter).
  - **Energy Properties:** Energy (kinetic, potential, internal), enthalpy, entropy, heat transfer rate, work done, power
  - **Momentum Properties:** Momentum, the force exerted on the control volume (force), pressure, impulse
  - **Mass Properties:** Mass, mass flow rate, mass flux, density
  - **Other:** Volume, volume flow rate

2. Surrounding Space

Surrounding space: The context or environment in which an object or system exists.

- **Local:** Referring to a small, immediate area or scale.
- **Global:** This refers to a large encompassing scale or the entire system.
  - **Spatial Distribution:** Pressure gradient, temperature gradient, concentration gradient, force field
  - **Interfacial Properties:** Heat transfer coefficient, mass transfer coefficient, stress, shear stress, surface tension

- **Geometric Properties:** Length, area, shape, surface area, volume of the control volume

3. **Surrounding Time**

   Surrounding Time: The temporal context or environment in which an event or process occurs.

   ➢ **Local Time:**
   - Refers to a specific point or instant in time.
   - Often associated with a particular location or event.

   Examples: Time of day, reaction time, event duration.

   ➢ **Global Time:**
   - Encompasses a larger timeframe or the entire duration of a process or system.
   - Often used for comparisons or trends.

   Examples: Historical period, lifespan, project timeline.

   ➢ **Temporal Distribution:**
   - Describes how properties or quantities change over time.
   - Analogous to spatial distribution but in the time dimension.

Examples: Temperature variation, velocity profile, concentration change, force variation.

- **Temporal Properties:**

  - Characteristics of time itself or its measurement.
  - Examples: Time interval, frequency, period, rate, time constant.
- ❖ **Note 1:** Unlike space, which has dimensions (length, width, height), time is typically considered one-dimensional.
- ❖ **Note 2:** While there's no direct equivalent to "interfacial properties" or "geometric properties" in time, the concepts of "temporal gradient" (rate of change) and "temporal scale" (characteristic time) can be considered analogous.

4. Behavior

Behavior: The action or reaction of the system or its surroundings.

- **Static:** Unchanging over time.
  - **Statistical Properties:** Mean, variance, standard deviation of properties, correlation

- **Dynamic:** Changing over time.

- **Dynamic Properties**: Stability, oscillation, damping, resonance

> **Condition**: The state or circumstances under which a system exists.
> - **Qualitative Properties**: Patterns, trends, anomalies

5. **Condition**

Condition: The state or circumstances under which a system exists.

- **Thermodynamic Properties**: Temperature, pressure, specific volume, internal energy
- **Mechanical Properties**: Stress, strain, elasticity, conductivity, viscosity, density
- **Material Properties**: Density, conductivity, specific heat capacity
- **Chemical Properties**: Concentration, reaction rate, equilibrium constant

Examples:

- **Energy**: Control volume and condition
- **Force**: Control volume, surrounding space, and condition
- **Acceleration**: Surrounding space and surrounding time

- **Pressure:** Control volume, surrounding space, and condition
- **Velocity:** Surrounding space and surrounding time
- **Mass flow rate:** Control volume and surrounding time
- **Mass:** Control volume and condition
- **Pressure gradient:** Surrounding space
- **Volume:** Control volume and condition
- **Temperature:** Control volume and condition
- **Number of Moles:** Control volume and condition
- **Oscillation frequency:** Surrounding time and behavior
- **Viscosity:** Condition and control volume
- **Time:** Surrounding time

Notes on Classification

1. Some properties can be classified in multiple ways depending on the context. For example, temperature can be a global property in a climate analysis.
2. Many properties are influenced by conditions, such as pressure, temperature, and composition.
3. The distinction between static and dynamic can be relative. Some properties might appear static over a short timescale but dynamic over a longer timescale.

Additional Considerations

- **Spatial properties**: Properties related to the distribution of something in space (e.g., population density, concentration).
- **Temporal properties**: Properties related to changes over time (e.g., growth rate, decay rate).
- **Statistical properties**: Properties derived from statistical analysis (e.g., mean, standard deviation, correlation).
- **Interdependencies**: Properties often belong to multiple categories. For instance, force can be a result of interactions within a control volume (internal forces) or with the surrounding space (external forces).
- **Units**: Consistent use of units is crucial for accurate calculations and comparisons.
- **Dimensions**: Understanding the dimensions of properties (e.g., length, mass, time) can help in property analysis and conversions.
- **Scale**: The choice of control volume influences property categorization. A microscopic control volume might reveal properties not apparent at a macroscopic level.

Step 3: Specifying the Reference Identity

How do we identify the Reference Identity within a phenomenon?

The Reference Identity is a composite term encompassing three essential components:

- **Reference Substance**: The primary material, substance, or entity involved in the phenomenon (e.g., solid, fluid, gas, plasma).
- **Reference Quantity**: The primary quantity or measurement aspect of the phenomenon (e.g., energy, pressure, temperature, velocity).
- **Reference Unit**: the standard used to quantify the reference quantity (e.g., Kelvin, Joule, Pascal, meters per second).

We can categorize a phenomenon using its Reference Identity as follows:

A. **Type A phenomena** directly involve the reference substance and indirectly relate to the corresponding reference quantity and unit. Examples include gas behavior (gas as reference substance), photon behavior (photon as reference substance), and turbulent incompressible fluid behavior (incompressible fluid as reference substance, such as water or blood).

B. **Type B phenomena** primarily focus on the reference quantity, with indirect implications for the corresponding reference unit and substance. Examples include the acceleration formula (with acceleration as the reference quantity), power formula (with power as the reference quantity), density formula (density as the reference quantity), and gravitational force (with gravitational force as the reference quantity).

C. **Type C phenomena** primarily focus on the reference unit, with indirect implications for the corresponding reference quantity and substance. Examples include phenomena related to Newton's laws (Newton as reference unit) and the Joule effect (Joule as reference unit).

This categorization provides a framework for understanding the nature of derived properties and how they relate to different physical systems. It can be useful in analyzing and modeling complex phenomena.

Step 4: Analyzing Energy Interactions

Quantify energy through its interactions with measurable system characteristics.

More descriptive: Determining energy by analyzing the interactions between the system's energy and each operational measurable property[1]. It implies that we can

---

[1] An operational measurable property is a quantifiable characteristic of a system that can be subject to mathematical operations.

represent energy in terms of each operational measurable property, used in the definition of the system, within a control volume defined for the problem at hand.

More technical: Quantifying energy based on its correlation with measurable system parameters.

Step 5: Finalizing the Outcome Equation

Develop a mathematical model accurately representing the problem and its solution.

More descriptive: Construct a precise mathematical formula that aligns with problem constraints and yields the desired outcome.

More technical: Deriving a mathematical equation (ordinary, differential, or partial differential equation) that accurately predicts the target variable within the problem's parameter space.

Step 6: Modifying the Equation (if required)

Validate the equation through comparison with real-world or experimental data.

More descriptive: Verifying the equation's accuracy by contrasting its predictions with empirical observations.

More technical: Assessing the obtained mathematical equation validity by correlating model outputs with experimental results.

Applications

This expanded framework can be applied to a wider range of fields, including:

- **Mechanics:** Analyzing forces, motion, and energy
- **Thermodynamics:** Studying energy transfer and system behavior.
- **Fluid mechanics:** Investigating fluid flow and pressure distribution.
- **Solid mechanics:** Analyzing stress, strain, and material properties.
- **Chemical engineering:** Modeling reactors, separators, and transport processes.
- **Other Engineering Fields**

By incorporating fundamental physical quantities into the categorization, we can gain a more comprehensive understanding of complex systems and processes.

❖ **Note:** Many properties can fit into multiple classifications depending on the specific context. For example, temperature can be both a local and global property, depending on the scale of analysis.

CHAPTER 1
CATEGORY A

Category A systems refer to phenomena that directly regard the reference substance.

By considering energy as a fundamental reference quantity, we can utilize energy or its derived units to identify key properties relevant to the reference identity, such as temperature or heat flux in a temperature control system.

Energy's influence on system behavior allows us to control these properties directly or indirectly by manipulating the energy input and output values.

We can control and modify the system's behavior by manipulating energy input and output. This can lead to significant changes in both the system's input/output values and the state of the reference substance (e.g., heating a material to raise its temperature).

Energy the Reference Quantity

Energy, as the reference quantity, can represent:

- **A direct energy term or quantity:** A measurable property with units of energy, such as kinetic energy or potential energy.

$$[E] = kg * m^2/s^2.$$

- **A derived energy term:** A product of terms or quantities whose combined dimensions equate to energy, like force multiplied by distance (work).

$$[E] = [P] * [V] = kg * m^2/s^2,$$
$$[E] = \left[\frac{1}{2}m * v^2\right] = kg * m^2/s^2.$$

- **A composite energy term:** A sum of products of terms, where each product represents an energy component, and their total dimensions equal energy.

$$[E] = \left[\frac{1}{2}mv^2\right] + [PV] + \cdots = kg * m^2/s^2.$$

The energy-based approach offers potential advantages like simplicity, efficiency, and scalability for complex systems.

Step 1: Defining the Problem

By considering the ideal gas as a system and employing energy as a fundamental reference quantity, we can analyze the interactions between measurable properties within a control volume, surrounding space, surrounding time, and conditions.

- **Control volume:** A fixed region in space chosen for analysis.
- **Surrounding space:** The space outside the control volume.
- **Surrounding time:** The time frame related to the phenomenon.
- **Condition:** Properties intrinsic to the system or its surroundings.
- **Behavior:** The action or reaction of the system or its surroundings.

Hence, we need to develop an equation describing the behavior of an ideal gas within a control volume energy due to the definition of energy E in terms of a control volume CV as a measurable property that corresponds to a pressure P and temperature T for the number of moles of gas particles.

This approach enables us to identify key properties influencing energy, and subsequently derive an equation describing ideal gas behavior.

CATEGORY A

Step 2: Identifying Measurable Properties

In controllable terms, the input and output values of measurable properties must be adjustable. This allows us to

- Induce significant changes in the system (input/output values, reference substance, or both) by manipulating these values.

- Identify the state of the reference substance based on the measured property values across different phases (subcritical, critical, supercritical).

Thus, we can employ the following measurable properties whose classifications are related to the problem at hand. So, we have:

- **Energy:** Control volume and condition

- **Pressure:** Surrounding space and surrounding time

- **Volume:** Control volume and condition

- **Temperature:** Control volume and condition

- **Number of Moles:** Control volume and condition

Considering the provided category, Energy, Pressure, Volume, Temperature, and Number of Moles should be analyzed using control volume analysis due to its interactions with the surroundings and internal conditions.

- **Energy** should be categorized under control volume and condition because
    - **Control Volume:** Energy is a property associated with the matter or system within the control volume.
    - **Condition:** The amount or type of energy (kinetic, potential, internal, etc.) depend on the state or condition of the matter within the control volume.
- **Pressure** should be categorized under control volume and condition because:
    - **Control Volume:** Pressure is the force exerted per unit area on a surface, reflecting the interaction with its surroundings.
    - **Surrounding Space:** Pressure is a measure of the force distributed over a given space.
    - **Condition:** Pressure is a property of matter within a system that depends on the condition of the system, reflecting the distribution of molecular kinetic and potential energy.
- **Volume** should be categorized under control volume and condition because:

- **Control Volume:** Volume is a property of the space defined by the control volume.

- **Condition:** Volume can change depending on the conditions within the control volume.

- **Temperature** should be categorized under control volume and condition because:

  - **Control volume:** Temperature is a property of the matter within a specific control volume.

  - **Condition:** Temperature is a system condition representing the average kinetic energy of its constituent particles.

- **Number of Moles** should be categorized under control volume and condition because:

  - **Control Volume:** The number of moles is a property of the matter contained within a specific control volume.

  - **Condition:** The number of moles is dependent on the specific state or condition of the matter within the control volume.

❖ **Note:** We do not have a time frame related to the phenomenon.

Step 3: Identifying the Reference Identity

Our analysis reveals that energy, expressed in its base unit or derived forms, can serve as a unifying unit across categories (A, B, and C). This is because energy allows us to control and quantify input and output values for any phenomenon. By manipulating energy, we can induce significant changes in a system's state.

When considering the behavior of an ideal gas behavior of a system, the following reference identity is established:

- **The Reference Substance:** Ideal Gas.

- **The Reference Quantity:** Energy.

- **The Reference Unit:** Energy, denoted as E, with the unit of $[E] = kg * m^2 / s^2$.

- ❖ **Note:** In the absence of a directly defined reference quantity within a phenomenon, energy and its associated units can serve as the default reference for identifying relevant measurable properties.

Step 4: Analyzing Energy Interactions

To describe ideal gas behavior, we focus on measurable properties that interact with energy, our chosen reference quantity. For a gas system, energy can be represented in terms of properties such as pressure, volume, temperature, and the number of moles, which relate to the system's control volume, surrounding space, time, and condition.

Then, we have the interactions between the system's energy and each operational measurable property. It implies that we can represent energy in terms of each operational measurable property, used in the definition of the system, within a control volume defined for the problem at hand.

❖ Note: Within a control volume, each property is measurable and contributes to defining energy, our reference quantity. Energy's unit serves as a consistent measure for the interactions among these properties. For ideal gas behavior, we focus on properties directly related to energy exchanges within the control volume.

To derive the equation describing ideal gas behavior within a control volume (**CV**), we analyze how energy, our reference quantity, interacts with key measurable properties. This involves expressing energy in terms of these properties while maintaining consistent energy units. These interactions encompass:

- The interaction between energy E with a volume V, a controllable property within the system:

$$[E_V] = [V] * [X] = m^3 * \frac{kg}{m * s^2} = \frac{kg * m^2}{s^2},$$

with,

$$[X] = \frac{kg}{m * s^2} = [P],$$

$$[V] = m^3,$$

represent energy $E_{V,P}$ in terms of both volume V and pressure P of the gas.

- The interaction between energy E, temperature T, and the number of moles n of gas particles:

$$[E_{T,n}] = [T] * [n] * [Y] = K * M * \frac{kg * m^2}{K * M * s^2}$$

$$= \frac{kg * m^2}{s^2}.$$

Therefore,

$$[Y] = \frac{kg * m^2}{K * M * s^2} = [R],$$

which leads to representing energy $E_{T,n,R}$ in terms of the number of moles n, temperature T, and another measurable property R within the control volume.

❖ It is important to note that for fluid systems, velocity and related quantities primarily arise from inlet and outlet conditions within the control volume. As such, they are not primary properties in this context.

Step 5: Finalizing the Outcome Equation

This problem involves a single set of measurable properties,

$$[C_S] = \frac{[Q_r]_{LHS}}{[Q_m]_{RHS}} = 1$$

$$C_S = \frac{Q_{r,LHS}}{Q_{m,RHS}} = 1,$$

to control input and output values. The resulting equation will balance these properties, with left and right sides representing equal quantities, where $Q_{r,LHS}$ and $Q_{m,RHS}$ represent the reference quantity on the left-hand side and the measurable quantity on the right-hand side of the outcome equation, respectively.

> In our example, we establish the following relationship:

$$[C_S] = \frac{[Q_r]_{LHS}}{[Q_m]_{RHS}} = \frac{[E_{LHS}]}{[E_{RHS}]} = \frac{[E_{V,P}]}{[E_{T,n,R}]} = \frac{[V]*[P]}{[T]*[n]*[R]} = 1.$$

Simplifying this expression, we find that C_S equals the ratio of $E_{V,P}$ to $E_{T,n,R}$:

$$C_S = \frac{E_{V,P}}{E_{T,n,R}} = \frac{V*P}{T*n*R} = 1.$$

This equality implies that E_{LHS} and E_{RHS} are equivalent, leading to the equation:

$$V * P = T * n * R.$$

This equation models the behavior of an ideal gas. A fundamental characteristic of this equation is the requirement for consistent energy units on both sides to accurately represent the gas's properties.

Step 6: Modifying the Equation (if required)

- ❖ **Note:** The subsequent sub-steps (6-1 to 6-3) are provided for reference and may not be applicable in all situations.

Step 6-1: Equation Verification

It is essential to compare the derived equation with the expected or target equation to confirm that the desired modifications have been accurately implemented. This cross-verification step helps ensure that the equation aligns with the intended outcome.

Step 6-2: Equation Testing and Troubleshooting

Upon deriving the equation, rigorous testing is imperative to assess its accuracy and reliability. A comprehensive evaluation should involve:

- Employing diverse input values to validate the equation's performance across various scenarios and conditions. If inconsistencies or errors arise, a meticulous review of the equation's components and logical structure is necessary.

- Applying the equation to real-world data or experimental results to assess its practical applicability. Numerical solutions can be employed to compare the equation's predictions with empirical observations.

❖ **Note:** If experimental data corroborates the theoretical predictions, the equation is considered validated. However, discrepancies may necessitate the inclusion of additional variables to accurately capture the underlying physical phenomena.

Step 6-3: Continuous Equation Refinement

To maintain the equation's accuracy and relevance, ongoing review and modification are crucial. Regular assessment of the equation's performance and alignment with new information or data is essential.

Step 6-4: Equation Documentation and Commentary

Effective documentation is vital for preserving the equation's value. A clear and concise description of the equation's purpose, underlying assumptions, and potential future applications should be included. Comprehensive documentation facilitates understanding and maintenance by both the original author and subsequent users.

CHAPTER 2
CATEGORY B

Category B refers to the phenomena that directly introduce the reference quantity. So, we have the introduced reference quantity and consider its corresponding unit as the reference unit.

Step 1: Defining the Problem

- ➤ Considering the acceleration formula as the phenomenon in question, we will leverage the concept of energy and its interaction with acceleration (the phenomenon we're interested in) to derive an equation for an accelerating object.

Leveraging the previously defined classifications of control volume, surrounding space, time, behavior, and conditions, we have established the following classifications:

- **Control volume:** A fixed region in space chosen for analysis.

- **Surrounding space:** The space outside the control volume.

- **Surrounding time:** The time frame related to the phenomenon.

- **Condition:** Properties intrinsic to the system or its surroundings.

- **Behavior:** The action or reaction of the system or its surroundings.

Step 2: Identifying Measurable Properties

We can identify relevant measurable properties for our analysis.

- **Energy:** Control volume and condition

- **Acceleration:** Surrounding space and surrounding time

- **Velocity:** Surrounding space and surrounding time

- **Time:** surrounding time

According to the provided category,

- **Energy** should be categorized under control volume and condition because

- **Control Volume:** Energy is a property associated with the matter or system within the control volume.

- **Condition:** The amount or type of energy (kinetic, potential, internal, etc.) depends on the state or condition of the matter within the control volume.

- **Acceleration** should be categorized under surrounding space and surrounding time because:

    - **Surrounding Space:** Acceleration describes a change in velocity over a distance.

    - **Surrounding Time:** Acceleration represents a change in velocity over time.

- **Velocity** should be categorized under surrounding space and surrounding time because:

    - **Surrounding Space:** Velocity describes motion relative to a reference point.

    - **Surrounding Time:** Velocity represents a change in position over time.

- **Time** should be categorized under surrounding time because:

- **Control Volume:** Time measures rates of change, frequency, and flow.

This analysis employs energy as a universal reference quantity and explores its interactions (relationships) with acceleration. By expressing energy in terms of acceleration, we aim to identify relevant measurable properties within a defined control volume, considering spatial and temporal factors, system behavior, conditions, and derived units.

Benefits: This approach provides a systematic way to derive the desired equation by focusing on energy interactions and their connection to the measurable properties of the accelerating object.

Step 3: Identifying the Reference Identity

Energy serves as a fundamental unifying concept across diverse physical phenomena (categories A, B, and C). This enables us to:

1. **Control and quantify input/output values:** We can adjust energy flow to influence a system's behavior.
2. **Alter the system's state:** By manipulating energy, we can significantly change a system's properties.

> Considering deriving a formula for acceleration as the problem at hand, the Reference Identity related to an accelerating object is as follows:
- **Reference Substance**: Any state of matter (solid, fluid, gas, or plasma).
- **Reference Quantity**: Energy E (universal reference quantity) and acceleration a (specific reference quantity).
- **Reference Unit**: The unit of acceleration $[a] = m/s^2$.

To derive an acceleration formula, we focus on measurable properties interacting with energy within a defined control volume (**CV**). This approach ensures energy unit consistency throughout the analysis.

Step 4: Analyzing Energy Interactions

We follow a specific procedure incorporating the interactions between energy (as the reference quantity) and each operational measurable property to derive the equation for accelerating an object within a control volume.

Measurable properties critical to Category B phenomena are those with significant influence on the reference substance, reference quantity, or reference unit. These properties have quantifiable values essential for controlling changes in these core elements.

> Within a defined control volume, the interaction between the system's energy E and acceleration a (the reference quantity) serves as the primary focus. This interaction (relationship) allows us to express energy in terms of the reference quantity a (as an operational measurable property) within the system. Thus, we have:

- Examining the interactions between energy and acceleration (the reference quantity) within the control volume yields:

$$[E_a] = [a] * kg * m = m/s^2 * kg * m$$
$$= m/s * kg * m/s.$$

Consequently,

$$[a] * kg * m = m/s * kg * m/s,$$

leading to

$$[a] = m/s * 1/s = [v] * \left[\frac{1}{t}\right].$$

Here, $[v] = m/s$ and $[1/t] = 1/s$ represent measurable properties whose input and output values can be controlled to influence acceleration significantly.

Combining these elements, we obtain:

$$[a] = [v] * \left[\frac{1}{t}\right].$$

Having identified these measurable properties and their corresponding units, we can proceed to construct equations relating them.

Step 5: Finalizing the Outcome Differential Equation

We can conserve energy on both sides of an equation by applying relevant measurable properties to the phenomenon.

This involves expressing energy either directly or as a product of other properties.

By adhering to the law of energy conservation, we can construct a formula where units are consistent on both sides.

> ➤ Considering *a* as the reference quantity, we express energy as the product of acceleration and an unknown term, ΔX, to represent other relevant properties.
>
> To maintain energy equivalence on both sides of the equation:

1. **Left hand-side (LHS):** We convert the energy unit $(kg * m^2/s^2) = Joule$ into a product of the acceleration unit $[a] = (m/s^2)$ and units associated with ΔX, $[\Delta X] = (kg * m)$,

$$[E_{LHS}] = [a] * [\Delta X] = (kg * m) * \left(\frac{m}{s^2}\right) = \left(\frac{kg * m^2}{s^2}\right),$$

2. **Right hand-side (RHS):** Using the definition of acceleration as the rate of change in velocity (from v_{t_i} to $v_{t_{i+1}}$) over time (from t_i to t_{i+1}), we rewrite the energy dimension in terms of velocity $[\Delta v] = (m/s)$, time $[1/\Delta t] = (1/s)$, and $[\Delta X] = (kg * m)$ units, ensuring dimensional consistency,

$$[E_{RHS}] = [\Delta X] * [\Delta v] * \left[\frac{1}{\Delta t}\right] = (kg * m) * \left(\frac{m}{s}\right) * \left(\frac{1}{s}\right).$$

❖ Note that velocity (v) and time inverse $(1/t)$ are the fundamental measurable properties associated with the acceleration unit $[a] = (m/s^2)$.

Defined as the rate of change in velocity from v_{t_i} to $v_{t_{i+1}}$) over time (from t_i to t_{i+1}), Acceleration expresses a product of velocity (v) and time inverse $(1/t)$.

To proceed, we must express the reference quantity as a product of its constituent measurable properties. The

CATEGORY B

resulting expression's dimensions must match those of the reference quantity.

> To balance energy input and output in the acceleration equation, we consider:

$$E_{\text{LHS}} = a * \Delta X,$$

$$E_{\text{RHS}} = \Delta v * \frac{1}{\Delta t} * \Delta X,$$

where:

- Δv represents the change in velocity, calculated as:

$$[\Delta v] = [v_{t_{i+1}} - v_{t_i}] = \frac{m}{s},$$

with units of meters per second (m/s).

- $1/\Delta t$ represents the inverse of the change in time, calculated as:

$$\left[\frac{1}{\Delta t}\right] = \frac{1}{[t_{i+1} - t_i]} = \frac{1}{s},$$

with units of inverse seconds ($1/s$).

We identify a singular set of measurable properties, denoted as:

$$[C_S] = \frac{[Q_r]_{LHS}}{[Q_m]_{RHS}} = 1$$

$$C_S = \frac{Q_{r,LHS}}{Q_{m,RHS}} = 1.$$

These properties influence the energy input and output on both sides of the outcome equation, $Q_{r,LHS}$ and $Q_{m,RHS}$.

- In our specific example,

$$[C_S] = \frac{[Q_r]_{LHS}}{[Q_m]_{RHS}} = \frac{[E_{LHS}]}{[E_{RHS}]} = \frac{[a] * [\Delta X]}{[\Delta v] * \left[\frac{1}{\Delta t}\right] * [\Delta X]} = \frac{[a]}{[\Delta v] * \left[\frac{1}{\Delta t}\right]}$$
$$= 1.$$

This equation demonstrates the equivalence of E_{LHS} and E_{RHS}, with consistent energy unit dimensions, culminating in the acceleration formula:

$$[a] = [\Delta v] * \left[\frac{1}{\Delta t}\right],$$

$$a = \frac{\Delta v}{\Delta t} = \frac{dv}{dt}.$$

Step 6: Document the Obtained Differential Equation

- ❖ **Note:** Steps 6-1 through 6-3 are included for reference, although they may not apply to every scenario.

Thorough documentation significantly improves the equation's usability and maintainability for both the original author and subsequent users.

CHAPTER 3
CATEGORY C

Category C encompasses phenomena that inherently introduces a reference unit. This allows us to directly identify the corresponding reference quantity.

Step 1: Defining the Problem

> ➢ Considering Newton's second law, we can utilize energy to quantify the force required to accelerate an object, which is a second-order differential equation. This involves establishing a relationship between energy, force, and the resulting change in motion over time (acceleration).

Based on previous definitions, we have established the following classifications:

- **Control volume:** A fixed region in space chosen for analysis.

- **Surrounding space:** The space outside the control volume.

- **Surrounding time:** The time frame related to the phenomenon.

- **Condition:** Properties intrinsic to the system or its surroundings.

- **Behavior:** The action or reaction of the system or its surroundings.

Step 2: Identifying Measurable Properties

We can utilize the following measurable properties, categorized based on their relevance to the problem:

- **Energy:** Control volume and condition

- **Force:** Control volume, surrounding space, and condition

- **Acceleration:** Surrounding space and surrounding time

- **Mass:** Control volume and condition

According to this category,

- **Energy** should be categorized under control volume and condition because

- **Control Volume:** Energy is a property associated with the matter or system within the control volume.

- **Condition:** The amount or type of energy (kinetic, potential, internal, etc.) depends on the state or condition of the matter within the control volume.

- Force should be categorized under control volume, surrounding space, and condition because:

  - **Control Volume:** Force acts on a specific volume, object, or system.

  - **Surrounding Space:** Force has a direction in space, which captures the dimensional aspect of force as a vector quantity.

  - **Condition:** Force can be subject to the influence of conditions like various factors in terms of temperature, pressure, or other environmental conditions.

- Acceleration should be categorized under surrounding space and surrounding time because:

  - **Surrounding Space:** Acceleration describes a change in velocity over a distance.

- **Surrounding Time:** Acceleration represents a change in velocity over time.

- **Mass** should be categorized under control volume and condition because:

  - **Control Volume:** Mass is a property of a specific volume or object.

  - **Condition:** Mass is a property that can be subject to the influence of conditions (e.g. relativistic effects).

This analysis employs energy as a universal reference quantity and observe its interactions (relationship) with force as the reference quantity to express energy in terms of force.

By expressing energy in terms of force, we aim to identify relevant measurable properties within the control volume based on their associated derived units.

Step 3: Identifying the Reference Identity

We have established that using the energy unit for category C as a reference allows us to identify relevant measurable properties related to the Reference Identity and the problem at hand.

➤ Given Newton's Second Law as our focus, we aim to quantify the relationship between applied force and an object's change in velocity over time (acceleration).

To comprehend Newton's Second Law, we require a definition of energy directly proportional to the force applied to an object causing change in its movement with respect to time.

This necessitates establishing an energy reference standard. Thus, the Reference Identity to preserve energy is as follows:

- The Reference Substance can be any state of matter: solid, liquid, gas, or plasma.
- Energy serves as the universal reference quantity, with force as the specific reference quantity.
- The Reference Unit would be Newton, the SI unit of force equivalent to Joule per meter ($J/m = kg * m/s^2$).
- Mass and acceleration, in conjunction with energy, define force (the reference quantity) within a control volume. These properties interact with energy to establish a force-energy relationship, ensuring consistent energy units across the system.

Step 4: Analyzing Energy Interactions

CATEGORY C

We have learned that to express Newton's second law as an equation, the following conditions are considered when constructing the resulting formula:

> By examining the interaction between energy and force, we can express energy in terms of force (as an operational measurable property). Additionally, by analyzing the interaction (relationship) between force (as the reference quantity) and other measurable properties, we can describe force using these properties, maintaining consistent energy units across the system. Therefore,

$$[E_F] = [F] * m = kg * m/s^2 * m$$
$$[F] * m = kg * m/s^2 * m$$
$$[F] = kg * m/s^2,$$
$$= [M] * [a],$$

It implies that we can describe energy in terms of force $[F] = kg * m/s^2$, mass $[M] = kg$, and acceleration $[a] = m/s^2$, as measurable properties.

So, the obtained parts together read as follows:

$$[E_F] = [F] * m = [M] * [a] * m,$$

and consequently, we have:

$$[F] = [M] * [a].$$

Step 5: Finalizing the Outcome Equation

The preceding steps demonstrate that this problem involves a single set of interrelated terms or measurable properties,

$$[C_S] = \frac{[Q_r]_{LHS}}{[Q_m]_{RHS}} = 1$$

$$C_S = \frac{Q_{r,LHS}}{Q_{m,RHS}} = 1.$$

Our goal is to manage the input and output values of these properties, where $Q_{r,LHS}$ and $Q_{m,RHS}$ represent the two sides of the resulting equation.

> In our example, we establish the following relationship:

$$[C_S] = \frac{[Q_{r,LHS}]}{[Q_{m,RHS}]} = \frac{[F]}{[M] * [a]} = 1.$$

Consequently, C_S equals the ratio of $Q_{r,LHS}$ to $Q_{m,RHS}$:

$$C_S = \frac{Q_{r,LHS}}{Q_{m,RHS}} = \frac{F}{M * a} = 1.$$

This equation implies that $Q_{E,r,LHS}$ and $Q_{E,r,RHS}$ are equivalent, leading to an outcome equation identical to Newton's second law of motion:

$$F = M * a.$$

Step 6: Document your equation and comment

❖ Note: Steps 6-1 through 6-3 do not apply to this specific example but include a reference for our study.

This documentation should detail the equation's purpose, the assumptions it relies on, and its potential future uses or modifications for clarity and adaptability.

CHAPTER 4

MATHEMATICAL MODELING SOME PROBLEMS

Now, put your skills to the test! Employ the described process to address the following problems. Detailed solutions can be found in the next chapter. Attempt these on your own initially.

1. Mathematical Modeling an Equation for the behavior of the Steady-State Flow Process.

2. Mathematical Modeling an Equation for the behavior of the Thermal Energy of a System.

3. Mathematical Modeling a Differential Equation for the Shaft Work of a System.

MATHEMATICAL MODELING OF STEADY-STATE FLOW EQUATION

Introduction to Steady-State Flow Process

The steady-state flow process is a fundamental concept in engineering and thermodynamics. It refers to a state where the conditions at every point in the system remain constant with time.

It implies that all rates within the apparatus are constant and there is no accumulation of material or energy within the system.

This process is essential for various industrial processes and allows optimal efficiency and control.

The term "steady state" can apply to any physical system, whether as a chemical reactor, heat exchanger, or car engine.

For each case, it describes a state where the system has reached a stable equilibrium and maintains that state as long as the conditions remain unchanged.

This state contrasts with a transient state, where the system is yet adjusting and changes over time.

A steady state flow process follows certain conditions. Primarily, all rates within the system must be constant.

It includes mass flow rates, energy transfer, and any chemical reactions involved in the process. If there is any variation in these rates, the system will not be in a steady-state process.

Another crucial condition is that there should be no accumulation of material or energy within the system.

It means that the amount of material entering the system must be equivalent to the amount leaving it. Any imbalance in these quantities will prevent the system from reaching a steady state.

The steady-state flow process is analogous to a bathtub with a running faucet and an open drain. When the inflow from the faucet matches the outflow through the drain, the water level in the tub remains stable.

It is a similar approach, considering a steady-state flow process, where the rates of material and energy entering and leaving the system are balanced.

One of the key benefits of a steady state flow process is that it allows for efficient and predictable operation of industrial processes.

By maintaining constant conditions of the system, engineers can optimize the system's functionality and achieve the desired output.

It also allows for easier control and regulation of the process, as any changes in the system can be quickly identified and addressed.

In summary, the steady state flow process considers a state in which all conditions within a system remain constant over time.

It requires constant rates and no accumulation of material or energy within the system.

Problem

Given the general case of a steady-state-flow process shown in Figure 1. Identify an equation for describing and solving this problem.

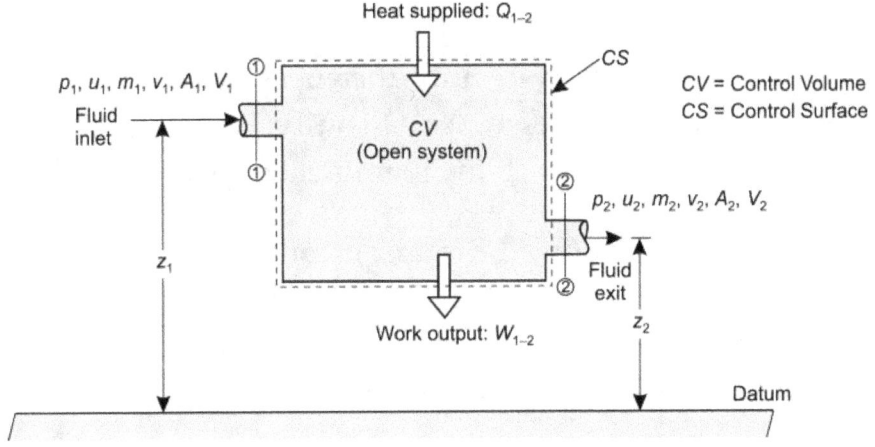

Figure 1. A schematic control volume (CV) for a steady-state flow process[2].

Solution:

For category A, energy (measured in Joules) serves as a universal reference quantity. By manipulating energy, we can control and modify the system's input and output properties.

Step 1: Defining the Problem

---

[2] A system with energy values in terms of work, heat, velocity, enthalpy, volume (specific volume includes), gravitational force, and mass rate.

- Kumar, S. (2022). Application of First Law of Thermodynamics to Flow Processes Thermodynamics. In: Thermal Engineering Volume 1. Springer, Cham. https://doi.org/10.1007/978-3-030-67274-4_4

Approach:

We will utilize energy (or its derived units) as the reference quantity to identify the measurable properties relevant to the Reference Identity. This approach capitalizes on energy's ability to influence system behavior.

By manipulating the input and output values of these properties, we directly control the system's energy, leading to significant changes in both the input/output values and the state of the reference substance.

As previously defined, we have established the following:

- **Control volume:** A fixed region in space selected for analysis.

- **Surrounding space:** The area external to the control volume.

- **Surrounding time:** The temporal scope of the phenomenon.

- **Condition:** Intrinsic properties of the system or its environment.

- **Behavior:** The actions or responses of the system or its environment.

We aim to develop an equation describing the steady-state flow process within a control volume.

Considering input or supplied heat (Q_{In}), output work (W_{Out}), and internal energy (E_{Int}), we will analyze the of a liquid mass with specified velocities (v_1, v_2), pressures (P_1, P_2) and volumes (V_1, V_2) at inlet and outlet, respectively. The system operates at different heights (Z_1, Z_2), as illustrated in Figure 1.

This analysis employs energy as a reference quantity to examine its interactions with each operational measurable property and their associated derived units within the control volume. Our goal is to express energy in terms of force.

By expressing energy in terms of supplied heat (Q_{In}), output work (W_{Out}), and internal energy (E_{Int}), we aim to identify the Reference Identity.

Step 2: Identifying Measurable Properties

So, we have: Therefore, we can utilize the following measurable properties relevant to the problem. These are:

- **Heat:** Control volume and condition
- **Work:** Control volume and condition

A STEADY-STATE FLOW PROCESS

- **Internal Energy:** Control volume and condition
- **Acceleration:** Surrounding space and surrounding time
- **Pressure:** Control volume, surrounding space, and condition
- **Velocity:** Surrounding space and surrounding time
- **Distance or Height:** Surrounding space
- **Volume:** Control volume and condition

According to this category,

- **Heat** should be categorized under control volume and condition because
    - **Control Volume:** Heat represents energy transfer across the system boundary.
    - **Condition:** Heat affects the state or condition of the matter within the control volume.
- **Work** should be categorized under control volume and condition because:
    - **Control Volume:** Work represents energy transfer across the system boundary.

- **Condition:** Work can change the state or condition of the matter within the control volume.

- **Internal Energy** should be categorized under control volume and condition because:

  - **Control Volume:** Internal Energy is a property of the matter within the system.

  - **Condition:** The total energy of the particles within the control volume.

- **Acceleration** should be categorized under surrounding space and surrounding time because:

  - **Surrounding Space:** Acceleration describes a change in velocity over a distance.

  - **Surrounding Time:** Acceleration represents a change in velocity over time.

- **Pressure** should be categorized under control volume, surrounding space, and condition because:

  - **Control Volume:** Pressure is a property of the substance within the system.

  - **Surrounding Space:** Pressure is a force exerted on a surface, which is a spatial concept.

- **Condition:** Pressure describes a state of the matter within the control volume.

- **Velocity** should be categorized under surrounding space and surrounding time because:
  - **Surrounding Space:** Velocity describes motion relative to a reference point.
  - **Surrounding Time:** Velocity represents a change in position over time.

- **Distance or Height** should be categorized under surrounding space because:
  - **Surrounding Space:** Distance represents a space relative to a reference point.

- **Volume** should be categorized under control volume and condition because:
  - **Control Volume:** Volume defines the physical space occupied by the system.
  - **Condition:** Volume is a property that describes the state of matter within the control volume.

Step 3: Identifying the Reference Identity

Our analysis reveals that energy, expressed in its base unit or derived forms, can serve as a unifying unit across categories (A, B, and C). This is because energy allows us to control and quantify input and output values for any phenomenon. By manipulating energy, we can induce significant changes in a system's state.

Specifically, in steady-state flow processes, energy (in the forms of heat and work) and its corresponding unit become the reference quantity and reference unit, respectively. This approach facilitates the identification of relevant measurable properties related to both the reference substance and the problem at hand.

➢ Considering the steady-state flow process as the primary phenomenon depicted in Figure 1, the following reference parameters apply:

- **The Reference Substance:** A fluid, such as water.
- **The Reference Quantity:** Energy, encompassing heat Q_{in}, work W_{out}, and internal energy E_{Int}.
- **The Reference Unit:** The unit of energy $[E] = kg * m^2/s^2$.

To describe steady-state flow, we examine measurable properties interacting with energy. Energy E is defined by input heat Q_{in}, output work W_{out}, and internal energy E_{Int} for liquid masses M_1, M_2 with varying velocities v_1, v_2, pressures P_1, P_2, volumes V_1, V_2, and elevations Z_1, Z_2 at the control volume inlet and outlet, as illustrated in Figure 1.

❖ Each property is measurable within the control volume **CV** depicted in Figure 1. These properties interact with energy to define the reference quantity (energy) within the system. To ensure consistency, all properties and energy are expressed in SI Units.

Step 4: Analyzing Energy Interactions

To derive the equation describing a steady-state flow process within a control volume, we follow a specific procedure that incorporates the interactions between energy (as the reference quantity) and each operational measurable property.

➢ By examining the interactions between energy (as the reference quantity) and each measurable property within the Control Volume **CV** of the system, we can characterize energy in terms of each operational measurable property. The unit of energy remains consistent across all these interactions:

- The interaction of internal energy E_{Int} with velocity v as an operational measurable property within the control volume shown in Fig. 1,

$$[E_{Int,1}] = [v] \cdot \frac{kg * m}{s} = \frac{kg * m^2}{s^2} = J,$$

$$= [v] * kg * \frac{m}{s} = [v^2] * kg,$$
$$= [v^2] * kg = [v^2] * [M] = [E_{v,M}],$$
$$[v^2] = \frac{m^2}{s^2},$$
$$[M] = kg.$$

This equation represents internal energy E_v in terms of both the velocity and mass of the fluid.

- The interaction of internal energy E_{Int} with pressure P as the operational measurable property within the control volume,

$$[E_{Int,2}] = [P] * m^3 = \frac{kg * m^2}{s^2} = J,$$
$$= [P] * [V] = [E_{P,V}],$$
$$[P] = \frac{kg}{m * s^2},$$
$$[V] = m^3.$$

This equation expresses internal energy $E_{Int,1}$ in terms of pressure P and volume V of the fluid.

- The interaction of internal energy E_{Int} with the mass M and elevation Z of the liquid within the control volume as follows:

$$[E_{Int,3}] = [Z] * \frac{kg * m}{s^2} = \frac{kg * m^2}{s^2} = J,$$
$$= [Z] * kg * \frac{m}{s^2} = J,$$
$$= [Z] * [M] * [g] = [E_{Z,M,g}],$$
$$[Z] = m,$$
$$[M] = kg,$$
$$[g] = \frac{m}{s^2}.$$

This equation expresses internal energy E_{Int} in terms of elevation Z, mass M, and gravitational acceleration g of the fluid.

Combining the previous results, we obtain:

$$[E_{Int}] = [E_{Int,1}] + [E_{Int,2}] + [E_{Int,3}].$$

Step 5: Finalizing the Outcome Equation

The problem involves a single set of interrelated terms or measurable properties. To balance the input and output values, we introduce a consistency factor (C_S):

$$[C_S] = \frac{[Q_r]_{LHS}}{[Q_m]_{RHS}} = \frac{[E_{LHS}]}{[E_{RHS}]} = 1,$$

$$C_S = \frac{(Q_{r,1} + \cdots + Q_{r,i})_{LHS}}{(Q_{m,1} + \cdots + Q_{m,j})_{RHS}} = 1,$$

where $(Q_{r,1} + \cdots + Q_{r,i})_{LHS}$ and $(Q_{m,1} + \cdots + Q_{m,j})_{RHS}$ represent the left and right sides of the equation, respectively.

> In our example, we establish the following relationship:

$$[C_S] = \frac{[Q_r]_{LHS}}{[Q_m]_{RHS}} = \frac{[E_{LHS}]}{[E_{RHS}]} = 1.$$

Consequently, C_S equals the ratio of E_{LHS} to E_{RHS}:

$$C_S = \frac{E_{LHS}}{E_{RHS}}$$

$$= \frac{(E_{v,M} + E_{P,V} + E_{Z,M,g})_{outlet} - (E_{v,M} + E_{P,V} + E_{Z,M,g})_{inlet}}{Q_{in} - W_{out}}$$

$$= \frac{(M_2 v_2^2 - M_1 v_1^2) + (P_2 V_2 - P_1 V_2) + g*(M_2 Z_2 - M_1 Z_1)}{Q_{in} - W_{out}}$$

$$= 1.$$

This equation indicates that E_{LHS} and E_{RHS} are equivalent. As a result, we derive the following energy balance equation for the steady-state flow process depicted in Figure 1:

$$(M_2 v_2^2 - M_1 v_1^2) + (P_2 V_2 - P_1 V_2) + g*(M_2 Z_2 - M_1 Z_1)$$
$$= Q_{in} - W_{out},$$

Crucially, all terms within this equation possess units of energy.

Step 6: Modifying and Verifying the Accuracy of the Equation

To verify the accuracy of the modified equation, we can conduct a thorough comparison between it and the original equation.

The presented methodology enables the derivation of mechanical energy from the foundational principles of Newton's second law of motion.

- **Starting with Newton's Second Law:** We employ Newton's second law, expressed as $F(x) = Ma$, as the fundamental equation to adapt the equation derived for the steady flow process. It is crucial to recognize that the force acting on the system can vary with position, represented by $F(x)$. Consequently, the equation becomes:

$$Ma - F(x) = 0.$$

Establishing a Connection Between Kinematics and Dynamics: By recognizing the inherent relationships among acceleration (a), velocity (v), and position (x), we can bridge the gap between kinematics and dynamics. Recognizing the fundamental relationship between acceleration and velocity, where acceleration is defined as the time rate of change of velocity ($a = \Delta v/\Delta t = dv/dt$), we can express the previous equation in terms of velocity. This yields the equation:

$$M\frac{dv}{dt} - F(x) = 0.$$

- **Deriving the Work-Energy Equation:** To progress towards the work-energy equation, we multiply both sides of Newton's second law by velocity (v) and apply the product rule of calculus. This manipulation yields the equation

$$Mv\frac{dv}{dt} - F(x)v = 0.$$

The product rule of calculus dictates that the derivative of the square of velocity with respect to time is

$$\frac{dv^2}{dt} = 2v\frac{dv}{dt}.$$

Multiplying both sides of this equation by the mass M, it reads:

$$M\frac{dv^2}{dt} = 2Mv\frac{dv}{dt}.$$

Continuing the derivation, we obtain:

$$\frac{d}{dt}\left(\frac{1}{2}Mv^2\right) = Mv\frac{dv}{dt}.$$

This step elucidates the origin of the one-half factor. Substituting this expression back into Newton's second law leads to the equation:

$$\frac{d}{dt}\left(\frac{1}{2}Mv^2\right) - F(x)v = Mv\frac{dv}{dt} - F(x)v = 0$$

- **Introducing Potential Energy:** To establish a connection between force and an object's position, we introduce the concept of potential energy, denoted as $V(x)$. A fundamental relationship exists between force and potential energy, expressed as

$$F(x) = -\frac{dV(x)}{dx},$$

where the force is equal to the negative derivative of potential energy with respect to position.

- **Using the Chain Rule:** By applying the chain rule, we can elucidate the relationship between the rate of change of potential energy and the force acting upon the object. This relationship is mathematically represented as

$$\frac{dV(x)}{dt} = \frac{dV}{dx} * \frac{dx}{dt} = -F(x)v.$$

- **Unifying Energy Components and Defining Mechanical Energy:** Combining the previously derived kinetic energy term ($Mv^2/2$) with the potential energy term, we can demonstrate that their sum remains constant over time.

This invariant quantity is termed total mechanical energy, denoted as E, and its time derivative is zero ($dE/dt = 0$). Mathematically, this is expressed as

$$\frac{d}{dt}\left(\frac{1}{2}Mv^2\right) + \frac{dV(x)}{dt} = \frac{d}{dt}\left(\frac{1}{2}Mv^2 + V(x)\right) = 0.$$

As a result, the sum of kinetic energy ($Mv^2/2$) and potential energy $V(x)$ remains unchanged over time. This constant value is recognized as the total mechanical energy of the system, denoted as $E_{k,p}$. Mathematically, integrating both sides of the equation with respect to time leads to the expression

$$E_{k,p} = \left(\frac{1}{2}Mv^2 + V(x)\right).$$

Furthermore, potential energy, represented by $V(x)$, can be described as:

$$V(x) = g * (M_2 Z_2 - M_1 Z_1).$$

In this equation, g signifies the acceleration due to gravity, while M_1, M_2, Z_1, and Z_2 correspond to relevant mass and height values.

Incorporating these energy terms into the derived equation results in the modified form:

$$\Delta E = \Delta H + \frac{1}{2}(M_2 v_2^2 - M_1 v_1^2) + g * (M_2 Z_2 - M_1 Z_1)$$

$$= Q - W_s.$$

Here, ΔH represents the change in enthalpy, calculated as:

$$\Delta H = (U_2 - U_1) + (P_2 V_2 - P_1 V_1),$$

where U represents internal energy, P represents pressure, and V represents volume.

Once you have derived your formula, it is crucial to test its validity in terms of accuracy. After building your formula, evaluate it thoroughly by

- Using different input values to ensure it produces accurate results to verify its accuracy in all scenarios before moving on to more complex calculations. (Note: If you encounter any errors or unexpected results, go carefully back through each step and component of the formula to review and double-check your logic and syntax), and

- Putting the obtained formula to the test, using practically reliable options such as numerical solutions compatible with as many theoretical and practical experiments and results as possible.

- ❖ Note: If the experimental results match the theoretical ones, you are on the right track. If not, you need to call in other terms (with more than one variable) based on which physical descriptions the phenomenon you are dealing with relies upon terms such as convection, diffusion, surface tension, gravity, etc. that represent multiple variables corresponding to one another.

Once you have a working formula, it is crucial to include comments, explaining the purpose of the formula, any assumptions made, and how it should be in the future (in terms of use or modification).

MATHEMATICAL MODELING OF HEAT TRANSFER EQUATION

Introduction to Thermal Heat Transfer

Thermal energy is a core concept in thermodynamics that examines the interplay between heat, work, and energy within a system.

It is often referred to as heat energy and is directly connected to the kinetic energy of the particles within a system.

This energy significantly influences a system's temperature, behavior, and various physical processes such as heat transfer and phase transitions.

The random motion of particles within a system generates thermal energy. The Boltzmann constant (**k**) is a crucial constant that links thermal energy to temperature.

This constant indicates that higher temperatures are associated with increased average kinetic and thermal energy.

Thermal energy constitutes a core component of thermodynamics, representing the internal energy of a substance.

What is Heat Transfer?

Heat transfer is the discipline concerned with the generation, use, conversion, and exchange of thermal energy (heat) between physical systems. Essentially, it is the study of how heat moves from one place to another.

Importance of Heat Transfer

Understanding heat transfer is crucial in various fields, including:

- **Engineering:** Designing efficient heat exchangers, insulation systems, and thermal management for electronic devices.
- **Building science:** Developing energy-efficient buildings by controlling heat gain and loss.
- **Climate science:** Studying the Earth's climate and the impact of greenhouse gases.
- **Medicine:** Understanding heat transfer in the human body for medical treatments and diagnostics.

Mathematical Representation

Heat transfer phenomena can be described mathematically using equations that represent the conservation of energy, momentum, and mass. These equations, combined with constitutive laws, form the basis for analyzing and predicting heat transfer processes.

Problem

Given the general form of thermal energy of a system shown in Figure 2. Identify an equation that is applicable for describing and solving this problem.

Solution

Step 1: Defining the Problem

As shown in Fig. 2, we need to develop an equation to describe heat transfer from object 1 with the temperature T_1 and mass M_1 to object 2, with the temperature T_2 and mass M_2, such that

$$T_1 > T_2,$$

to change the temperature T_2 of object 2.

Leveraging energy's profound influence on system behavior, we will utilize energy (or its derived units) as the reference quantity to identify the measurable properties relevant to the Reference Identity.

Through manipulation of the input and output values of the measurable properties, we exert direct control over the system's energy. This manipulation induces significant changes in both the input/output values and the state of the reference substance.

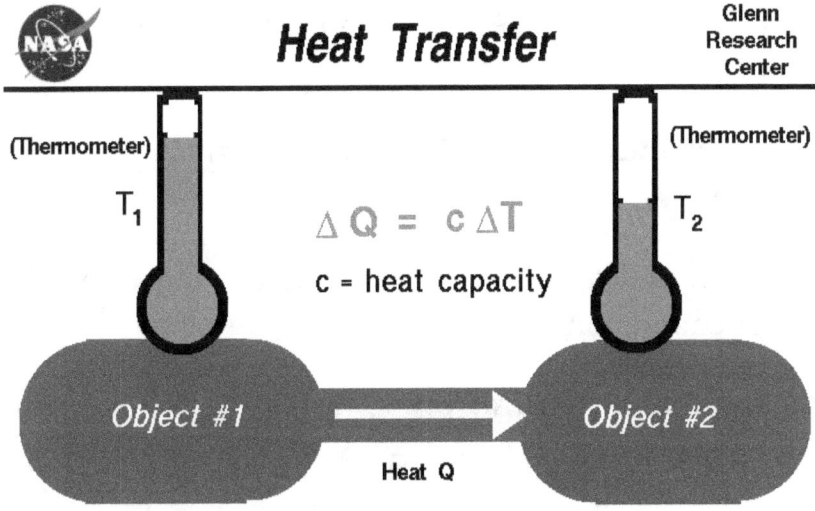

Figure 2. A schematic representation of a thermal heat transfer system as a phenomenon in question[3].

---

[3] Thermodynamics investigates the energy exchanges and transformations within a system.

- NASA. (n.d.). *Heat transfer*. NASA. https://www.grc.nasa.gov/www/k-12/airplane/heat.html

Step 2: Identifying Measurable Properties

As previously defined, the following definitions provide a foundation for analyzing a system:

- **Control volume:** A fixed region in space selected for analysis.

- **Surrounding space:** The area external to the control volume.

- **Surrounding time:** The temporal scope of the phenomenon.

- **Condition:** Intrinsic properties of the system or its environment.

- **Behavior:** The actions or responses of the system or its environment.

Thus, we can employ the following measurable properties whose classifications are related to the problem at hand. So, we have:

- **Heat:** Control volume and condition

- **Temperature:** Control volume and condition

- **Specific Heat Capacity:** Control Volume and Condition

Based on the provided category,

- **Heat** should be categorized under Control volume and condition because
    - **Control Volume:** Heat represents energy transfer across the system boundary.
    - **Condition:** Heat affects the state or condition of the matter within the control volume.
- **Temperature** should be categorized under control volume and condition because:
    - **Control Volume:** Temperature is a property of the matter within the system.
    - **Condition:** Temperature describes the thermal state of the matter within the control volume.
- **A Control Property of the Control Volume CV, in terms of heat capacity** most closely aligns with control volume and condition because:
    - **Control Volume:** Heat Capacity is a property of a substance within a defined system.

- **Condition:** Heat Capacity describes how a substance responds to changes in temperature within that control volume.

So, key measurable properties for describing heat transfer include:

1. Energy (in the form of Heat)

2. The amount of heat essential to change the temperature of 1.00 kilogram of mass by 1.00 degree as a measurable property for representing a control property of the control volume **CV**, in terms of heat, denoted by C_p.

3. Temperature **T**.

4. Mass **M**.

❖ Note: All properties are measured within the control volume (CV) depicted in Figure 2. These properties interact to define energy within the system, with all values expressed in SI units.

Step 3: Identifying the Reference Identity

We have determined that the unit of energy can serve as a unifying unit for all categories (A, B, and C) because it allows us to control and quantify input and output values for any given phenomenon. By manipulating energy, we can significantly alter the system's state.

For heat transfer analysis, energy (in the form of heat) and its associated unit serve as the reference quantity and reference unit, respectively. This framework helps identify relevant measurable properties for the problem.

- Considering thermal energy transfer from object 1 to object 2 (as depicted in Figure 2), the following reference parameters apply:
  - **The Reference Substance:** Solid, fluid, or gas.
  - **The Reference Quantity:** Energy in the form of heat Q.
  - **The Reference Unit:** Energy unit $[E] = kg * m^2 / s^2$.

To describe heat transfer, we focus on measurable properties interacting with energy. Specifically, consider heat transfer from object 1 (temperature T_1, mass M_1) to object 2 (temperature T_2, mass M_2). The resulting temperature change (ΔT) in object 2 allows us to express energy in terms of these properties. This ensures consistent energy units throughout the heat transfer process within the control volume (**CV**) depicted in Figure 2.

Step 4: Analyzing Energy Interactions

To derive the equation describing a steady-state flow process within a control volume, we follow a specific procedure that incorporates the interactions between energy (as the reference quantity) and each operational measurable property.

> By analyzing the interactions between energy (as the reference quantity) and each measurable property within the control volume, we can characterize energy in terms of these properties. Notably, the unit of energy remains consistent throughout these interactions:

- The interactions between energy (in the form of heat Q as the reference quantity) and temperature (as the operational measurable property), we have:

$$[Q] = [C_p] * [T] * kg = \frac{Joule}{kg * K} * K * kg = J = \frac{kg * m^2}{s^2}.$$

- The interactions between energy (in the form of heat Q as the reference quantity) and mass of an object undergoing heat transfer (as the operational measurable property), we have:

$$[Q] = [C_p] * [M] * K = \frac{Joule}{kg * K} * kg * K = J = \frac{kg * m^2}{s^2}.$$

So, the obtained parts together read as follows:

$$[Q] = [C_p] * [T] * [M].$$

Step 5: Finalizing the Outcome Equation

The problem involves a single set of interrelated terms or measurable properties. To balance input and output values, we introduce a consistency factor (**Cs**):

$$[C_S] = \frac{[Q_r]_{LHS}}{[Q_m]_{RHS}} = 1,$$

$$C_S = \frac{Q_{r,LHS}}{Q_{m,RHS}} = 1,$$

where $Q_{r,LHS}$ and $Q_{m,RHS}$ represent the left and right sides of the equation, respectively."

➢ As for our example, we have

$$[C_S] = \frac{[Q_r]_{LHS}}{[Q_m]_{RHS}} = \frac{[E_{LHS}]}{[E_{RHS}]} = \frac{[Q]}{[C_p]*[T]*[M]} = 1,$$

$$C_S = \frac{Q_{r,LHS}}{Q_{m,RHS}} = \frac{E_{LHS}}{E_{RHS}} = \frac{\Delta Q}{C_p * \Delta T * \Delta M}$$

$$= \frac{(Q_2 - Q_1)}{C_p * (T_2 - T_1) * (M_2 - M_1)} = 1.$$

It implies that $E_{LHS} = E_{RHS}$, resulting in the following equation for heat transfer, as illustrated in Figure 2:

$$\Delta Q = C_p * \Delta T * \Delta M.$$

This equation represents the heat transfer formula for the system.

Step 6: Document the Obtained Equation and Comment

❖ Note: We can skip steps 6-1 to 6-3 as they do not apply to our example.

Provide comments to clarify the purpose of the formula, outline any assumptions made, and describe its intended future use or potential modifications. This practice ensures that others (and your future self) can easily understand, apply, and maintain the formula over time.

MATHEMATICAL MODELING OF SHAFT WORK EQUATION OF A SYSTEM

Introduction to Shaft Work of a System

Shaft work is a form of mechanical energy exerted by a rotating shaft resulting from the torque applied to it and its angular displacement. The definition of shaft work is proportional to the force applied at an angle to a disc (mounted on a rigid shaft), causing displacement of the disc rotating around an axis (angular displacement) per unit of time over a period of time. It is a mechanism to transfer power between components within a machine.

The unit of shaft work in the International System of Units (SI) is the joule (J).

Shaft work has diverse applications, including:

- Power transmission in machines,
- Lifting and lowering loads,
- Rotating machinery (e.g., pumps, fans, compressors),
- Steering vehicles.

The efficiency of shaft work depends on friction between the shaft and its supporting bearings.

Friction converts a portion of shaft work into heat, potentially reducing machine efficiency.

Low-friction bearings and regular lubrication can enhance shaft work efficiency.

Examples of shaft work include:

- Crankshaft in a car engine, converting reciprocating piston motion into rotary motion for power transmission to wheels.
- Propeller shaft in a boat, converting engine rotation into thrust for propulsion.
- Driveshaft in a bicycle, transmitting power from pedals to the rear wheel for motion.

In summary, shaft work is a mechanical energy form exerted by a rotating shaft used for power transmission within machines.

Problem

Given the general case for the shaft work of a system shown in Figure 3. With its relevant measurable properties as variables. Identify an equation that is applicable for describing and solving this problem.

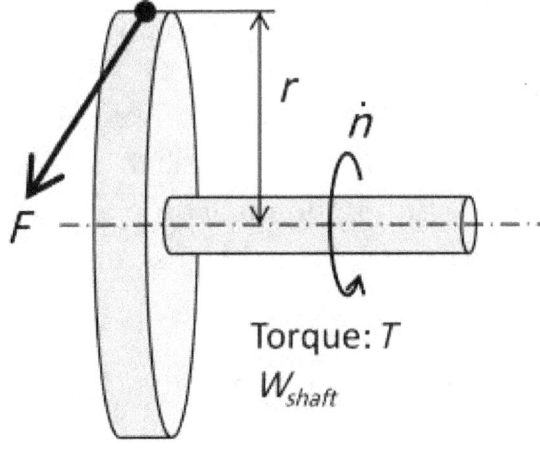

Figure 3. A schematic representation of shaft work considering rotational linear force and the number of revolutions of the shaft[4].

Solution

Step 1: Defining the Problem

Given the general case for the shaft work of a system shown in Figure 3, where the relevant measurable

---

[4] Beyond boundary work, various forms of work exist. One common type is shaft work, illustrated schematically in Figure 3. A constant force, F, acts tangentially along the rim of a disk. If the disk undergoes n rotations, the displacement is calculated as $2\pi r n$.

- *Shaft work*. Shaft Work - an overview | ScienceDirect Topics. (n.d.).
  https://www.sciencedirect.com/topics/engineering/shaft-work

properties are treated as variables, we need to identify an equation applicable for describing and solving this problem.

We aim to find a relationship for the amount of energy required to rotate the shaft at a specific angular velocity over a finite amount of time, as illustrated in Figure 3. The figure depicts a rotating disc attached to a shaft with an external force applied.

As previously defined, the following definitions provide a foundation for analyzing a system:

- **Control volume:** A fixed region in space selected for analysis.

- **Surrounding space:** The area external to the control volume.

- **Surrounding time:** The temporal scope of the phenomenon.

- **Condition:** Intrinsic properties of the system or its environment.

- **Behavior:** The actions or responses of the system or its environment.

Step 2: Identifying Measurable Properties

We can consider the following measurable properties relevant to the problem. So, we have:

- **Work:** Control volume and condition

- **Force:** Control volume, surrounding space, and condition

- **Number of Rotations per Unit of Time:** Surrounding space and surrounding time

- **Angular Displacement:** Surrounding space

- **Circumference of the Rotating Disc:** Surrounding space

- **Time:** Surrounding time

According to this category,

- **Work** should be categorized under control volume and condition because:
    - **Control Volume:** Work represents energy transfer across the system boundary.
    - **Condition:** Work can alter the state of the matter within the control volume.

- **Force** should be categorized under control volume, surrounding space, and condition because:

    - **Control Volume:** Force acts on a specific volume or object within the system.

    - **Surrounding Space:** Force has a direction in space and magnitude, making it a vector quantity.

    - **Condition:** Force can be subject to the influence of factors such as temperature, pressure, or other environmental conditions.

- **Number of Rotations per Unit of Time** should be categorized under surrounding space and surrounding time because:

    - **Surrounding Space:** Number of Rotations involves rotational motion, a spatial characteristic.

    - **Surrounding Time:** Number of Rotations explicitly includes a unit of time, indicating a rate of change.

- **Angular Displacement** should be categorized under control surrounding space because:

- o **Surrounding Space:** Angular displacement describes the change in orientation of an object relative to a reference point.

- **Circumference of the Rotating Disc** should be categorized under surrounding space because:

  - o **Surrounding Space:** Circumference of the Rotating Disc is a geometric property related to the shape and size of the disc.

- **Time** should be categorized under surrounding time because:

  - o **Surrounding time:** Time measures rates of change, frequency, and flow. It is a measure of duration and is independent of the system.

Step 3: Identifying the Reference Identity

By using the energy unit for category C as a reference, we can identify corresponding measurable properties relevant to the Reference Identity for the problem.

- ➤ To understand shaft work of a system W_S (as illustrated in Figure 3), consider a force applied tangentially to a disc (mounted on a rigid shaft) causing angular displacement over time. To preserve energy in this form, the following reference parameters apply:

- **The Reference Substance:** A solid shaft with dominant properties of viscosity and incompressibility (e.g., water, blood, oil, honey).
- **The Reference Quantity:** Energy, in the form of shaft work (W_S), is the primary quantity on the left-hand side (LHS) of the equation.
- **The Reference Unit:** Energy unit $[E] = kg * m^2 / s^2$.

To describe shaft work within a control volume (**CV**), we focus on measurable properties that interact with energy. A rotating disc with an attached shaft serves as a reference for these energy interactions. By expressing energy in terms of these properties and their interactions, we ensure consistent energy units throughout the (**CV**).

Therefore, the measurable properties essential to characterizing shaft work are as follows:

1. **Energy:** In the form of shaft work.

2. **Force:** Exerted on the system.

3. **Distance or angular displacement:**

    - **Circumference of a rotating disc:** Calculated as $S = 2\pi r$.

- **Number of rotations per unit time:** Represented by $\dot{n}$.
- **Time:** Required for the rotation motion t.

Each of these properties is quantifiable within a control volume (**CV**), which is defined as a rotating disc equipped with a shaft. This control volume serves as a reference framework for examining the interactions between energy and the identified measurable properties. By expressing energy in terms of these properties and their corresponding interactions, we ensure the consistency of the energy unit throughout the entire system within the control volume.

❖ Note that the units are in the International System of Units (SI).

Step 4: Analyzing Energy Interactions

To express shaft work as an equation, we can utilize the following principles in formulating the resulting expression:

➢ We can employ the relationships between energy (as the reference quantity) and each relevant measurable property within the control volume (**CV**). The predominant measurable property can serve as an operational measurable property for describing energy in terms of its characteristics, ensuring that the energy unit remains consistent across all interactions.

To examine the interaction between energy E and force F, with force as an operational measurable property, we establish the following relationship:

$$[E_F] = [F] * m = \frac{kg * m}{s^2} * m = \frac{kg * m^2}{s^2}.$$

To express force F in terms of torque T, we begin by partitioning a rotating disc into infinitesimal elements, each with a radius designated as dr.

Employing the circumference of such an element, denoted as $dS = 2\pi dr$, within the defined control volume (CV), we introduce a novel measurable quantity symbolized as $[\bar{O} * T]$ and equated to $[F]$, representing torque. Consequently, the derivative of torque with respect to the vector distance is expressed as:

$$[E_F] = [F] * m.$$

Furthermore, the equation

$$[F] * m = [\bar{O} * T] * [dS] = \left(\frac{kg * m}{s^2}\right) * m = \frac{kg * m^2}{s^2},$$

emerges. By applying the operator $\bar{O}$ to torque T, the resulting unit is determined as follows:

$$[F] = [\bar{O} * T] = \frac{kg * m}{s^2}.$$

Therefore, the operator $\bar{O}$ must correspond to the del operator, ∇, equivalent to d/dX, where X represents the vector, (x, y, z).

Applying this operator to torque yields

$$[\bar{O} * T] = \left[\frac{dT}{dX}\right] = [F],$$

ultimately resulting in the unit of force,

$$[F] = \left(kg * \frac{m}{s^2}\right).$$

➢ Analyzing the interaction between energy and the number of rotations (n), we convert it to rotations per unit time $(\dot{n})$ and time (t) to form an operational measurable property. We have:

$$[E_{\dot{n}}] = [n] * \left(\frac{kg * m^2}{s^2}\right) = [n] * [E_F] = \frac{kg * m^2}{s^2}$$

$$[E_{\dot{n}}] = [\dot{n}] * [t] * [E_F] = \frac{1}{s} * s * \left(\frac{kg * m^2}{s^2}\right),$$

$$[\dot{n}] * [t] = \frac{1}{s} * s.$$

Combining these results, we find:

$$[E_{\dot{n}}] = [\dot{n}] * [t] * [E_F] = [n] * [E_F].$$

Step 5: Finalizing the Outcome Equation

From the preceding steps, we observe that this problem involves a singular set of interconnected terms or measurable properties, expressed as:

$$[C_S] = \frac{[Q_r]_{LHS}}{[Q_m]_{RHS}} = 1,$$

$$C_S = \frac{Q_{r,LHS}}{Q_{m,RHS}} = 1.$$

This equation serves to regulate the input and output values of the constituent measurable properties, wherein $Q_{r,LHS}$ and $Q_{m,RHS}$ denote the respective terms on the left-hand side and right-hand side of the resulting equation.

➤ In our specific example, we obtain:

$$[C_S] = \frac{[Q_r]_{LHS}}{[Q_m]_{RHS}} = \frac{[E_{LHS}]}{[E_{RHS}]} = \frac{[W_S]}{[\dot{n}] * [t] * [E_F]} = \frac{[W_S]}{[2\pi r \dot{n} t\, F]} = 1,$$

$$C_S = \frac{E_{LHS}}{E_{RHS}} = \frac{W_S}{\dot{n} * t * E_F} = \frac{W_S}{2\pi r \dot{n} t\, F} = 1.$$

The equality of $E_{LHS} = E_{RHS}$ yields the following equation for the system's shaft work energy:

$$W_S = \dot{n} * t * E_F = 2\pi r \dot{n} t\, F.$$

Step 6: Document the Obtained Equation and Comment

❖ **Note:** Steps 6-1 to 6-3 are not applicable to this example.

This documentation should clearly articulate the equation's purpose, any underlying assumptions, and potential future applications or modifications.

CHAPTER 5
A HARD PROBLEM

Mathematical Modeling of a Partial Differential Equation for the Behavior of Viscous Incompressible Fluids on Three-Dimensional Space

Step 1: Defining the Problem

This chapter delves into the characteristics of viscous, incompressible fluids within a three-dimensional space, which refers to a system of partial differential equations. To comprehend the behavior of these fluids, it is essential to translate their properties into a mathematical framework. This necessitates a comprehensive understanding of the problem domain.

The focal point of this analysis is the behavior of viscous, incompressible fluids, with viscosity identified as the predominant measurable property and incompressibility as the primary condition.

1. **Viscosity:** It represents the internal resistance to shear forces generated between fluid layers moving at varying velocities. This resistance arises from the frictional interactions between liquid molecules.

2. **Incompressibility:** It signifies a control volume characterized by constant density and an absence of volume change.

Step 2: Incompressibility and the Measurable Properties

To quantify the characteristics of a system, it is necessary to analyze its control volume, spatial context, temporal domain, state, and actions to identify measurable properties. Building upon previously established definitions, the following terms provide a fundamental framework for system analysis:

- **Control volume:** A designated, stationary, and fixed region within space selected for analysis.
- **Surrounding space:** The area external to the control volume.
- **Surrounding time:** The temporal scope of the phenomenon under consideration.
- **Condition:** Intrinsic properties of the system or its environment.

- **Behavior:** The actions or responses of the system or its environment.

The measurable properties responsible for the behavior of viscous incompressible fluids are:

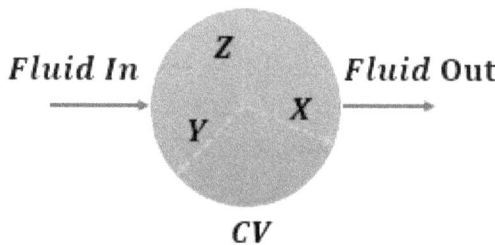

- **Energy:** Control volume and condition
- **Force:** Control volume, surrounding space, and condition
- **Pressure:** Control volume, surrounding space, and condition
- **Velocity:** Surrounding space and surrounding time
- **Viscosity:** Control volume and condition
- **Density:** Control volume and condition
- **Time:** Surrounding time

The reasoning behind this classification is as follows:

- **Energy** should be categorized under control volume and condition because
    - **Control Volume:** Energy is a property associated with the matter or system within the control volume.

- o **Condition:** The amount or type of energy (kinetic, potential, internal, etc.) depends on the state or condition of the matter within the control volume.
- **Force** should be categorized under control volume, surrounding space, and condition because:
  - o **Control Volume:** Force acts on a specific volume, object, or system.
  - o **Surrounding Space:** Force has a direction in space, which captures the dimensional aspect of force as a vector quantity.
  - o **Condition:** Force can be subject to the influence of conditions like various factors in terms of temperature, pressure, or other environmental conditions.
- **Pressure** should be categorized under control volume, surrounding space, and condition because:
  - o **Control Volume:** Pressure is a property of the substance within the system.
  - o **Surrounding Space:** Pressure is a force exerted on a surface, which is a spatial concept.
  - o **Condition:** Pressure describes a state of the matter within the control volume.
- **Velocity** should be categorized under surrounding space and surrounding time because:

- - Surrounding Space: Velocity describes motion relative to a reference point.
    - Surrounding Time: Velocity represents a change in position over time.
  - Viscosity should be categorized under control volume and condition because:
    - Control volume: Viscosity is a characteristic associated with the fluid within a specific boundary (the control volume) describes a change in velocity over a distance.
    - Condition: Viscosity is dependent on the state of the fluid within the control volume, primarily temperature and pressure.
  - Density should be categorized under control volume and condition because:
    - Control volume: Density is a characteristic associated with the matter contained within a specific boundary (the control volume).
    - Condition: Density is dependent on the state of the material within the control volume, such as temperature, pressure, and composition.
  - Time should be categorized under surrounding time because:
    - Surrounding time: Time measures rates of change, frequency, and flow.

Step 3: Identifying the Reference Identity

Given the focus on the behavior of viscous, incompressible fluids, we establish the following reference points:

- **The Reference Substance:** A viscous incompressible fluid with viscosity as a dominating measurable property and incompressibility as a dominating condition (e.g., water, blood, oil, honey, etc.).
- **The Reference Quantity:** The concept of a reference quantity, denoted as Q_r, is pivotal in this framework. It can represent a fundamental property such as energy, force, acceleration, or a derived quantity obtained through mathematical operations on measurable properties. Force, due to the significant influence of viscosity.
- **The Reference Unit:** The unit of force, expressed as $[F] = kg * m / s^2$, consistent with the International System of Units (SI).

Both reference quantities and operational measurable properties can be interconnected through mathematical equations to define energy. For example, energy can be expressed in terms of multiple operational measurable properties, such as force and acceleration, along with their respective units.

Step 4: Analyzing Energy Interactions

Each of the following properties can be quantified as an operational term in relation to the reference quantity (force) to describe the interactions between energy and force:

1. Energy
2. Force
3. Velocity
4. Pressure
5. Constant density
6. Viscosity

When energy is not the reference quantity, its interaction (relationship) with the reference quantity can be categorized as follows:

- Interactions between energy and force with energy as the dominant unit and force as the operational measurable property (assuming an external force).

- Interactions between energy and force with energy as the dominant unit and pressure as the operational measurable property.

- Interactions between energy and force with energy as the dominant unit and velocity as the operational measurable property.

- Interactions between energy and force with energy as the dominant unit and viscosity as the operational measurable property.

❖ It is important to note that the energy unit remains consistent throughout the control volume (**CV**), subject to the divergence-free condition of fluid flow:

$$Fluid_{In} = Fluid_{Out}.$$

It implies that the sum of changes in velocity with respect to the $x, y,$ and z axes within a control volume CV is zero as there is no net change in velocity (Fluid inflow equals fluid outflow).
Mathematically, this is represented as:

$$\nabla \cdot v = \frac{dv_x}{dx} + \frac{dv_y}{dy} + \frac{dv_z}{dz} = 0$$

Deriving the Navier-Stokes Equation: An Energy-based Approach

To describe the complex behavior of a viscous, incompressible fluid, we adopt an energy-based approach. This method involves analyzing the interactions between energy and its associated properties.

1. By examining the interaction between energy and force (as our reference quantity), we identify the dominant influence on energy, ensuring that energy is expressed in terms of this reference quantity through consistent units.

2. Analyzing the interaction (relationship) between the reference quantity and every measurable property within the system, we determine the dominant influence on the reference quantity, ensuring its representation in terms of each property through consistent units.

We begin by identifying force as the reference quantity that influences energy within the fluid system.

By ensuring that energy is consistently expressed in terms of force units, we establish an initial framework for subsequent analysis.

This involves determining how energy varies with changes in force and other relevant properties.

To comprehensively understand the system, we examine the relationship between the reference quantity (force) and every measurable property within the fluid.

By establishing how force is influenced by these properties, we ensure that the entire system is accounted for.

- For the initial analysis, we consider the interaction between energy E and force F. Here, energy is treated as the dominant quantity, while force serves as the reference quantity. Mathematically, this relationship can be expressed as:

$$[E_F] = [F] * m = \frac{kg * m^2}{s^2} * m = \frac{kg * m^2}{s^2}.$$

To facilitate analysis, we divide the fluid into infinitesimally small volumes, denoted as dV.

This approach allows for analyzing the interactions between energy and force within this infinitesimal control volume.

Let $\bar{O}$ represent an operator that acts upon elements within the three-dimensional space $\mathbb{R}^3$. The operational capabilities of $\bar{O}$ are versatile, encompassing a spectrum of mathematical processes including differential operators, gradient calculations, material derivatives, integration, and other relevant mathematical operations. This adaptability empowers $\bar{O}$ to interact appropriately with diverse quantities, executing corresponding operations precisely. A fundamental constraint imposed upon $\bar{O}$ is the preservation of energy unit consistency across the entire system of equations.

➢ In the context of fluid dynamics, Force can be expressed in terms of pressure, **P**, and a spatial operator, denoted as $\bar{O}$, to incorporate pressure into the interactions between energy and force while maintaining energy as the primary quantity.

This operator, when applied to pressure (the multiplication of operator $\bar{O}$ and **P**)[5], provides a measure of the pressure gradient within the fluid element. However, for accurate representation and consistency with standard mathematical practices, the gradient operator, **∇**, should be employed instead of the generic operator $\bar{O}$.

The gradient operator applied to pressure (**∇ * P**) yields the pressure gradient, which, when multiplied by the differential volume (**dV**) and a proportionality constant, represents the force acting on the element:

$$[E_{F,P}] = [F] * m$$

$$[F] * m = [dV] * [\bar{O} * P] * m = m^3 * \left(\frac{kg}{m^2 * s^2}\right) * m$$

$$= \frac{kg * m^2}{s^2}.$$

---

[5] The expression ($\bar{O} * P$) represents the product of an operational measurable property, $\bar{O}$, and pressure, **P**, both acting on the differential control volume, **dV**.

The quantity $\nabla * P$ represents the rate of change of pressure per unit length and has units of

$$[\bar{O} * P] = [\nabla * P] = \frac{kg}{m^2 * s^2} = \frac{1}{m} * \frac{kg}{m * s^2}.$$

➢ Force is also influenced by fluid density, ρ, and velocity, v. To incorporate velocity into the interactions between energy and force while maintaining energy as the primary quantity, the operational factor $\bar{O}$, when applied to velocity ($\bar{O} * v$), acts analogous to the material derivative operator $\bar{O} := D/Dt$, or equivalently ($v * \nabla$), to preserve energy as the primary unit.

➢ This operator accounts for both temporal and spatial changes in velocity and allows for the expression of force in terms of velocity, density, and spatial variations.

$$[E_{F,v}] = [F] * m.$$

By applying an operational factor $\bar{O}$ to velocity v, and considering both acting on a differential control volume dV, we can incorporate density ρ as an intrinsic fluid property that significantly influences velocity behavior. This leads to:

$$[F] * m = [dV] * [\rho] * [\bar{O} * v] * m = m^3 * \frac{kg}{m^3} * \left(\frac{m}{s^2}\right) * m$$
$$= \frac{kg * m^2}{s^2}.$$

From this equation, we derive:

$$[\bar{O} * v] = \frac{m}{s^2} = \frac{1}{s} * \frac{m}{s},$$

Expanding the material derivative operator in Cartesian coordinates yields:

$$\bar{O} := (v * \nabla) = v_x \frac{\partial}{\partial x} + v_y \frac{\partial}{\partial y} + v_z \frac{\partial}{\partial z}.$$

Consequently, we arrive at the expression:

$$\left[\frac{dv}{dt}\right] + \left[v * \frac{dv}{dX}\right] = \frac{1}{s} * \frac{m}{s}.$$

➢ By examining the interactions (relationship) between energy and force, with energy established as the fundamental unit, we can express force in terms of viscosity, a material property, by introducing an operational factor denoted as $\bar{O}$ and a corresponding measurable property, Q_m. This methodological approach preserves energy as the primary unit of measurement.

To quantify these interactions, dimensional analysis is employed. The energy associated with force and viscosity, $E_{F,\mu}$, is equated to the product of force, F, and a linear dimension, m. This leads to the expression:

$$[E_{F,\mu}] = [F] * m.$$

Expanding the force term in terms of volumetric flow rate, dV, viscosity, μ, the operational factor, $\bar{O}$, the measurable property, Q_m, and a linear dimension, m, yields:

$$[F] * m = [dV] * [\mu] * [\bar{O} * Q_m] * m$$

Substituting the base units for each quantity, we obtain:

$$m^3 * \frac{kg}{m*s} * \left(\frac{1}{m*s}\right) * m = \frac{kg*m^2}{s^2}.$$

To identify the appropriate forms for $\bar{O}$ and Q_m, we must scrutinize the available physical properties: energy, force, pressure, velocity, and viscosity. The objective is to find a combination that produces a quantity with units of $1/{m*s}$, corresponding to the product $\bar{O} * Q_m$.

Upon analysis, velocity, composed of meters and seconds, exhibits the closest alignment with the required units. Consequently, we assign Q_m as velocity, represented by v.

To determine the operational factor $\bar{O}$, we seek a mathematical operation that, when applied to velocity, yields units of $1/{m*s}$. Given that $\bar{O}$ is an operator acting on velocity, it must involve either distance or time. Considering the initial units of velocity m/s, and the target units $1/{m*s}$, we conclude that $\bar{O}$ is a differential operator applied twice with respect to distance. Mathematically, this is expressed as:

$$\bar{O} := \nabla^2.$$

The Operational Factor $\bar{O}$ (Physical Interpretation): Identifying the operational factor $\bar{O}$ as the Laplacian operator, ∇^2 while mathematically sound, a deeper physical interpretation is essential.

The Laplacian is often associated with curvature or spatial rate of change. In the context of fluid dynamics, it might represent the rate of change in velocity with respect to position.

It could also be linked to the concept of vorticity, which represents a measure of the local rotation of a fluid element.

Step 5: Finalizing the Outcome Partial Differential Equation

Building upon the preceding analysis, it becomes evident that the system under consideration is a collection of multiple, distinct reference quantity constituents. Each of these constituents engages in a reciprocal relationship with its corresponding measurable properties.

These properties can be mathematically represented as follows:

$$[C_S] = \frac{[Q_r]_{LHS}}{[Q_m]_{RHS}} = 1,$$

$$C_S = \frac{(Q_{r,1} + \cdots + Q_{r,i})_{LHS}}{(Q_{m,1} + \cdots + Q_{m,j})_{RHS}} = 1,$$

In these equations, C_S is a system constant. The term $Q_{r,LHS} = (Q_{r,1} + \cdots + Q_{r,i})_{LHS}$ represents the sum of reference quantity constituents on the left-hand side of the resulting equation. Similarly, $Q_{m,RHS} = (Q_{m,1} + \cdots + Q_{m,j})_{RHS}$ represents the sum of measurable properties corresponding to each reference quantity constituent. This equation establishes a fundamental equilibrium condition within the system describing fluid behavior, ensuring a balance between the input and output values of these properties.

➢ For describing fluid behavior, we have:

$$C_S = \frac{(Q_{r,1} + \cdots + Q_{r,i})_{LHS}}{(Q_{m,1} + \cdots + Q_{m,j})_{RHS}} = \frac{E_{LHS}}{E_{RHS}}$$

$$= \frac{F * m}{(\rho * (\partial_t * v + (v * \nabla) * v) + \nabla * p - \mu * \nabla^2 * v) * m}$$

$$= \frac{F}{\rho * (\partial_t * v + (v * \nabla) * v) + \nabla * p - \mu * \nabla^2 * v} = 1$$

❖ It is important to note that the negative sign in front of the viscous term, $(-\mu * \nabla^2 * v)$, indicates the opposing directions of flow and shear forces.

This equality, indicating $E_{LHS} = E_{RHS}$, leading to the final equation:

$$F = \rho * (\partial_t * v + (v * \nabla) * v) + \nabla * p - \mu * \nabla^2 * v.$$

This equation represents the Navier-Stokes equation for a 3D viscous incompressible fluid.

Step 6: Document the Obtained Equation and Comment

Include comments explaining the purpose of the formula, underlying assumptions, and potential future uses or modifications.

BIBLIOGRAPHY

- Cengel, Y. A., & Boles, M. A. (2006). *Thermodynamics: An engineering approach* (5th ed.). McGraw-Hill.

- Incropera, F. P., DeWitt, D. P., Bergman, T. L., & Lavine, A. S. (2011). Fundamentals of heat and mass transfer. Wiley.

- Bejan, A. (2013). Heat transfer. Wiley.

- Constantin, P., & Foias, C. (2020). *Navier-Stokes equations*. University of Chicago Press.

- Douglas, J. F., Gasiorek, J. M., & Swaffield, J. A. (2001). *Fluid mechanics*. Addison-Wesley.

- Hibbeler, R. C. (2020). *Engineering mechanics: Statics and dynamics* (14th ed.). Pearson.

- Serway, R. A., & Jewett, J. W. (2004). *Physics for scientists and engineers* (6th ed.). Brooks/Cole.

Acknowledgments

I extend my sincere appreciation and my utmost love and gratitude to my Mentor and Savior, Imam Al-Hojat Ibn Hassan Al-Askari, for the unwavering support, guidance, and wisdom provided during the challenging moments of my life.

With sincere love and thanks to my Father, Mother, and Brothers who stood by me and for their unwavering belief and encouragement.

To everyone who believed in this project, even when faced with moments of adversity.

www.ingramcontent.com/pod-product-compliance
Lightning Source LLC
Chambersburg PA
CBHW071036240526
45469CB00006BD/2224